WISLEY HANDBOOK 27

Fruit Pests
Diseases and Disorders

AUDREY BROOKS, KEITH HARRIS
and ANDREW HALSTEAD

LONDON
The Royal Horticultural Society
1976

Contents

Printed in England by
Henry Stone & Son (Printers) Ltd., Swan Close, Banbury, Oxfordshire

Pests, Diseases and Disorders

Amateur gardeners who wish to grow good crops of high quality fruit can do so, in spite of the pests and diseases which may affect their fruit at some time or other, by first learning how to recognise specific troubles from the symptoms which they produce and then to apply the correct control measures. In this book are described the most troublesome pests and diseases of fruit, and treatments for their control are given. The emphasis is on those troubles which reduce yields, damage the fruits or interfere with the growth of young plants.

Good health cannot be achieved unless good quality fruit trees, bushes and canes are planted and for the best results with soft fruits, whenever possible, plants certified to be healthy and true to type should be bought, as should virus-tested trees of top fruit when available. Planting should be carried out correctly and the plants well maintained.

Most fruits are grown as perennials and their permanence favours a build-up of pests and diseases. The grower can do a great deal to prevent such a build-up by good hygiene and cultural treatment and should, therefore, observe the following basic principles:

—control weeds and cut the grass regularly in and around fruit plantations to limit the amount of shelter to pests. Irrespective of whether the trees are grown in grass or under arable conditions, always maintain a clear area around the base of each tree.

—clean out hedge bottoms to limit the carry-over of pests and diseases from one year to the next.

—collect fallen leaves and fruits, especially those obviously diseased; remove dead shoots and mummified fruits. Do not put diseased material on the compost heap but burn it.

—use barriers and traps against certain pests such as strawberry beetle and winter moth.

—prune as necessary to prevent dense growth so allowing good air circulation, and burn the prunings to kill overwintering eggs of pests and fungal spores.

—avoid overcrowding of plants, especially soft fruits, to help prevent diseases such as grey mould and powdery mildew.

—keep the plants well fed and watered.

The use of cultural methods for pest and disease control will help to reduce the damage, but in any case, it is essential to give fruit crops the correct cultural treatment to produce good crops and to prevent the

3

occurrence of physiological disorders. Some of the more common troubles of this type are also described in this book, together with measures for their prevention.

Whilst cultural measures are important for the control of pests and diseases, in many instances, chemicals are also necessary and for some troubles, spraying is the best or only method of control. Amateur gardeners can adopt a routine spray programme, as commercial growers do, to obtain high quality fruit. Chemicals used properly are efficient, but they can be a danger if misused. The gardener should, therefore, use chemicals only when really necessary, selecting those suitable for the type of pest or disease to be controlled and applied at the correct times, according to the recommendations given in the following pages. It is essential to read the instructions on the manufacturer's label and to apply the chemical at the correct concentration.

It is also important to have efficient equipment for applying insecticides and fungicides. Soft fruits and trees on dwarfing rootstocks can be sprayed effectively with one of the many small hand or pressure sprayers on the market, of 1 or 2 gallon capacity (4.5–9.0 litres). These are usually made of plastic and are either pumped continuously while spraying, or are pumped up before spraying. This latter type is easier to use, especially when designed to be carried on the operator's back as a knapsack sprayer. With pressure sprayers, the pressure must be kept up during spraying otherwise control is liable to be inefficient.

Whilst small trees can be covered with the spray solution when using these sprayers, for taller trees an extension lance may be needed. On very large trees, especially standards, it is impossible to make a thorough application using anything other than a mechanised sprayer. Such machines are expensive and their purchase usually out of the question for small-scale fruit growing, and most gardeners have to accept poorer quality fruit from very large old trees. When buying new fruit trees for the garden, it is advisable to select those which have been grafted on dwarfing rootstocks, as these trees will not grow beyond a manageable height.

The best conditions for spraying are when it is dry, calm and frost-free. It is necessary to spray on a dry day as the foliage is then also dry and the solution will stick to it. There are very few days when it is completely calm, however, but spraying can be done when there is a light breeze, provided some arrangements are made to screen adjacent plants, for instance with plastic sheets, if the spray is likely to harm other plants. The spray should be applied from all sides of the plant to give as good a coverage as possible.

It is essential to avoid killing bees and other pollinating insects, so to minimise the risk, plants should not be sprayed with insecticides during flowering except in unavoidable situations, mentioned specifically later in this book, and then spraying should be done in the evening when the bees

4

are not working. It would also be helpful if gardeners notified local beekeepers of their intention to spray with insecticides, so that the hives could be closed.

When spraying make sure that the material is mixed thoroughly before applying the solution. Nozzles on fruit tree sprayers are designed to give a cone-shaped spray. The narrower the cone the further the distance the spray will carry but the area covered will be less. Wide angle cones are suitable for close work and narrow cones for tall trees or when the wind is strong. Some sprayers have a number of nozzles to suit the various conditions but others have a single adjustable nozzle which gives a range of spray patterns.

A coarse high pressure spray should be used when applying a winter wash and a fine to medium spray in the spring or summer when the objective is to obtain an even cover on both the upper and lower leaf surfaces. As a general guide, spray to run-off but no more than this otherwise the material is wasted.

Inefficient spraying can do more harm than good, by killing a pest's natural predators while having little effect on the pest itself. If there is a strain resistant to the insecticide or fungicide among the surviving population, this may then multiply unchecked.

Another point to remember in disease control is that most fungicides available to amateurs are not eradicants but protectants so that they prevent fungal infection and must be applied before the disease actually appears. Peach leaf curl, apple scab and grey mould are just some of the diseases which are controlled by applying a fungicide long before symptoms appear.

When using chemicals it is necessary to ensure that the fruit is not contaminated with harmful or distasteful residues. Some sprays such as BHC, malathion, captan and thiram may cause a taint in the fruit, which becomes more noticeable if it is preserved in some way, so other suitable chemicals should be used in preference if the fruit is to be bottled, deep frozen or used for jam. To prevent harmful residues on the fruit, a minimum interval must be observed between spraying and harvesting and the relevant information is always given on the manufacturer's label.

Any chemicals used in the garden must be handled carefully and stored safely out of reach of children and away from food. Spray equipment must be washed out carefully after use so that no traces of the chemical are left, and insecticides and fungicides should not be applied with apparatus which has been used for weedkillers. As a final note of precaution, it must be mentioned that great care must be taken to prevent contamination of ponds, ditches and waterways with chemicals or used containers.

In this book amateur gardeners will find a guide towards producing high quality fruit on well grown plants, which should be the aim of all good gardeners.

Apples

Aphids (Greenfly). There are at least five species of aphid which will attack apples, but only three of these are of any importance. The most damaging is the rosy apple aphid which infests the young shoots and causes the leaves to become yellow and curled. The aphids excrete honeydew, and the foliage becomes sticky and soiled with the sooty moulds that grow on this sugary substance. They also feed on the young fruitlets and this makes the mature fruit small and distorted.

The rosy leaf-curling aphid also distorts the leaves and turns them bright red. This is a very distinctive aphid but less common than the above. Both of these aphids overwinter on the tree as small shiny black eggs which are often laid near buds. They hatch in the spring and the aphids remain active on the tree until June-July, when the rosy apple aphid migrates to its summer host, plantains, and the rosy leaf-curling aphid lays its overwintering eggs and dies out. The rosy apple aphid migrates back to apples to lay its eggs in the autumn.

The other aphid commonly found on apple is the woolly aphid, sometimes known as "american blight". This lives in colonies on the branches and it sucks sap from the bark. Woolly aphids are covered by a white waxy material that can look like a mould. Their feeding causes the development of soft warty growths on the branches, which may split open and allow the entry of canker spores. This aphid overwinters on the tree as young aphids under loose flakes of bark.

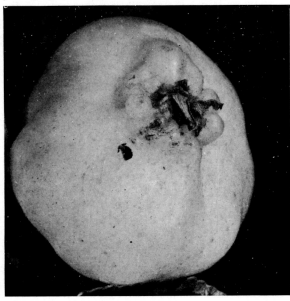

Fig. 1. Damage to apple fruits by rosy apple aphid.

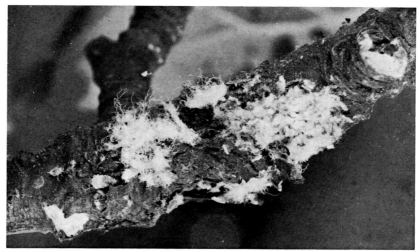

Fig. 2. A colony of woolly aphids on apple twigs showing the typical waxy threads excreted by the pest.

Treatment. A tar-oil winter wash thoroughly applied in December–January will kill overwintering eggs. This should be followed by a systemic insecticide such as dimethoate or formothion during the green cluster stage of bud development. Woolly aphid is not controlled by winter washes and they should be forcibly sprayed with dimethoate, formothion or malathion after petal fall when they become active.

Apple sucker. This pest is related to aphids and they damage the flower buds as they open. The blossoms are often discoloured and these symptoms may be mistaken for frost damage. Close examination will reveal the young suckers which look like flattened aphids. The overwintering eggs are pale yellow and shaped like miniature rugby footballs. Treatment is the same as for aphids.

Birds. Bird damage cannot be disregarded, and they are a major pest in some areas. Bullfinches and sparrows attack blossom buds of most fruit trees and bushes early in the season, and tits, blackbirds and starlings are among those that damage ripening fruit. However slight the fruit damage caused by a bird's beak, it is always a potential entry for disease. The initial wounds also attract wasps and other insects which make the damage worse.

Treatment. The only sure way of preventing damage by birds to buds and fruit is to use netting to exclude them. A permanent fruit cage is best in gardens adjoining woodland where birds are numerous but temporary

7

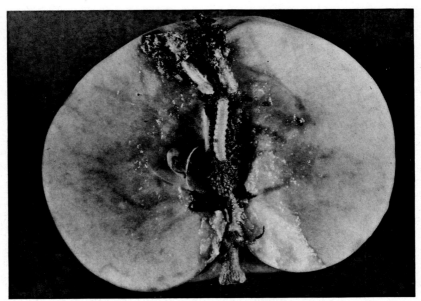

Fig. 3. Codling moth larva inside the apple.

netting may prove effective in some circumstances. If the smaller birds, such as the tits, are to be excluded, the mesh should be no larger than $\frac{3}{4}$in. (19mm) rigid or $\frac{1}{2}$in. (13 mm) pliable. Where netting is not possible individual fruits may be protected by enclosing them in paper, cellophane or polythene bags but these must have adequate ventilation.

Codling moth. The caterpillars of this moth tunnel into the fruits while they are developing, so ruining them for eating. The caterpillars overwinter in cocoons, in cracks and under tree ties. They pupate in May and moths emerge in May-June and in the next weeks lay eggs on leaves and fruitlets. The new generation of caterpillars soon hatch and tunnel into the developing fruits. In mid-August the caterpillars leave the fruits to find somewhere to overwinter.

Treatment. Chemical control of the moth is difficult, especially on large trees. Fenitrothion sprayed four weeks after petal-fall (about mid-June) with one or two repeat applications at fortnightly intervals will kill the young caterpillars before they enter the fruit.

Non-chemical control may be achieved by trapping the overwintering caterpillars. First scrape off any loose bark on the trunks and branches and then tie sacking or corrugated cardboard around them by mid-July.

8

Fig. 4. Damage by apple sawfly larvae, showing the typical ribbon-like scar.

The caterpillars spin cocoons in or under these bands and can then be killed by removing and burning the bands in the winter. This treatment will have no effect on the damage done to the fruit in the season when it is applied but may reduce infestations in subsequent seasons if it is used regularly on all apple trees in an area.

Apple sawfly. Apple sawfly caterpillars also burrow into fruit but, in contrast to codling caterpillars, feed on the surface of the fruits before entering. This type of sawfly damage can be recognised by the characteristic ribbon scar on the fruit skin at harvest. Apples which have been tunnelled by the caterpillars do not ripen and fall from the tree in June. The caterpillars overwinter in the soil as pupae and the small, inconspicuous adults emerge in the spring and fly up onto the blossom to lay eggs. The caterpillars tunnel into the core of the fruit, feed for about a month, and then leave the fruits to overwinter in the soil.

Treatment. Spraying with BHC, dimethoate or fenitrothion will control apply sawfly but the timing of this treatment is critical. Apply the spray about a week after blossoming, so that all the young fruitlets are thoroughly coated with the chemical, which acts on contact with the hatching caterpillars. Picking and destroying infested fruits in June will also help to limit later attacks.

Fig. 5.
Winter moth larvae
on apple leaves.

Caterpillars of the fruit tree tortrix, the vapourer, lackey moths and the winter moths also damage apple trees. The winter moths are usually most troublesome as they attack the buds, young leaves and blossom in early spring.

Winter moth caterpillars are of the "looper" type. They hatch from late April onwards and feed until May or June when they go into the soil until the following winter. Spraying with trichlorphon at the late green cluster stage will kill the young caterpillars. Grease bands placed round the trunks between October and March will trap the wingless females as they climb up the trunks into the trees to lay their eggs.

Fruit tree tortrix caterpillars will be largely controlled by sprays applied against codling moth. If other caterpillars are noticed on the foliage later in the year, trichlorphon can be used to control them.

Capsids. Both apple capsid and common green capsid attack apples. Signs of capsid feeding are punctured, tattered, puckered and distorted leaves. Damaged fruits develop characteristic bumps and superficial rough corky patches. Buds may sometimes be killed. Capsid bugs feed by sucking sap and the damage results from toxins in the capsids' saliva which are introduced into the plant as they feed.

Treatment. Capsids are not easily controlled as they tend to be rather elusive. Eggs overwintering on the trees may be killed by spraying thoroughly with DNOC/petroleum in January or February, but sprayed areas may be re-invaded during the spring and summer. It is generally better to delay control until after flowering. Spray at petal fall with malathion or with one of the systemic insecticides, such as formothion or dimethoate.

Fig. 6. Capsid damage on an apple fruit.

Fruit tree red spider mite. Red spider is not often a problem on unsprayed trees because natural enemies keep them at a low level. Regular use of insecticides may encourage red spider mite by killing their predators. The mites live in colonies on the underside of leaves and they suck sap from the leaf tissues. When mite populations are high affected leaves are discoloured and may eventually dry up, die, and fall prematurely. If the undersides of the leaves are examined with a magnifying glass from late June onwards, small reddish brown mites and their minute spherical eggs may be seen. This pest also attacks plum trees.

Treatment. Overwintering eggs can be killed by spraying thoroughly with DNOC/petroleum in February, but it is probably best to delay treatment until shortly after blossoming, when thorough spraying with malathion, formothion or dimethoate will check infestations, providing the mites have not acquired resistance to these materials. Spraying with dinocap to control apple mildew will also suppress red spider mite.

Canker is the most common and the most destructive disease of apple trees. Sunken and discoloured patches develop on the bark and as these

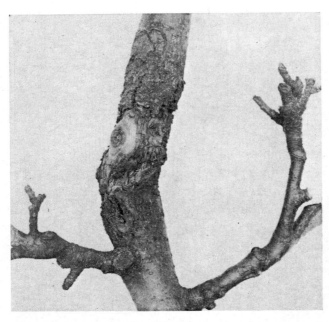

Fig. 7.
A bad infection
of apple canker
which has
encircled the
branch.

extend they become elliptical with the bark shrinking in concentric rings around each canker. In summer white pustules of fungus spores form on the sunken bark. The branch usually becomes swollen around the canker and if the shoot is completely girdled by the canker, the part above dies. The fungus overwinters as small red fruiting bodies which develop in the cracked and diseased bark. Infection through wounds, pruning cuts and leaf scars can occur at any time of the year.

Treatment. Diseased branches and spurs should be cut out and burnt. On branches where the cankers have not caused any dieback cut out the brown diseased tissues with a chisel or sharp knife and collect the parings. Paint the clean wounds with a canker paint.

Where the disease is severe and the trees are not too tall, it is worthwhile applying three sprays of bordeaux mixture or liquid copper starting after the fruit has been picked but before leaf fall. Give the second application when half the leaves have fallen, and the final one in spring as the buds burst. The spring application is the most important as most spores are released then.

Canker is more troublesome on trees growing in poorly drained soil, and the drainage should therefore be improved. This is particularly important for canker-susceptible cultivars such as 'Cox's Orange Pippin', 'Lord Suffield', 'James Grieve' and 'Ribston Pippin'. It is important to ensure that trees do not lose vigour through neglect.

Fig. 8. The terminal shoot of an apple lateral showing mildew infection.

Apple mildew is usually noticed in spring when the emerging shoots and flower trusses appear grey due to a mealy covering of fungus spores. Infected shoots are stunted, affected flowers do not set fruit and diseased leaves fall. The spores are released from May to September with the peak in mid-June when the new season's buds which are just developing become infected and remain so until they burst the following spring.

Treatment. Considerable control of mildew may be achieved by removing silvered shoots during pruning and cutting off any mildewed shoots as they are noticed in spring and later in summer when secondary infection has occurred. This secondary infection can, however, be prevented to a certain extent by mulching and watering as mildew is most troublesome on trees affected by drought.

Chemical control with the following materials is also possible:

Dinocap at pink bud stage (late April to early May), repeated at intervals of 7 to 14 days until mid-July.

Thiophanate-methyl at green cluster and then at 14-day intervals.

Benomyl may also give some control and lime sulphur may be used on cultivars which are not sensitive to sulphur.

A DNOC/petroleum winter spray applied for control of mite eggs will also help control mildew.

13

Fig. 9. Superficial scabs on the skin surface of apple fruits caused by apple scab.

Apple scab. Symptoms on the fruit are brown or blackish superficial scabs developing on the skin without internal rotting. In severe cases fruit entirely covered by scabs becomes mis-shapen and cracked. Olive green blotches develop on the leaves. They may fall prematurely, causing the tree to lose vigour with subsequent reduction in crop the following year. Small blister-like pimples develop on young shoots. Later these burst the bark and appear as ring-like cracks or scabs.

Treatment. Overwintering stages of the fungus should be removed by raking up and burning the fallen leaves in autumn and by cutting out cracked and scabby shoots when pruning. The disease can be controlled by spraying at regular intervals during the growing season with a protectant fungicide, but the fungus cannot be eradicated once it has become established.

Captan is the safest spray to use as it does not harm any cultivars, whereas lime sulphur, which also gives good control of the disease, must not be used on cultivars such as 'Stirling Castle' and 'St Cecilia' which are particularly sensitive to sulphur. Other cultivars ('Beauty of Bath', 'Rival', 'Belle de Boskoop', 'Lord Derby', 'Newton Wonder', Cox's Orange Pippin', 'Duchess's Favourite', 'Blenheim Orange' and 'Lane's Prince Albert') should not be sprayed with lime sulphur after blossoming. Whichever of these two fungicides is used, the first application should be given when the flower clusters are still tightly closed (green cluster), the second when the flower buds are showing pink (pink bud), the third when 80% of the petals have fallen and a fourth, three weeks after petal fall. A $2\frac{1}{2}\%$ solution of lime sulphur should be used at the green cluster and pink bud stages, and a 1% solution at petal fall and fruitlet stages. Lime sulphur must not be used after flowering if it has not been applied before flowering.

The new systemic fungicides benomyl and thiophanate-methyl will also give a good control of scab, but the first application must be given at bud burst and the trees should then be sprayed regularly once a fortnight. However, too frequent use of these fungicides can lead to the development of tolerant strains of the fungus.

The number of applications required for any of these fungicides depends on the severity of attack and in some seasons it may be necessary to continue spraying until late July.

Brown rot. Wounds in fruits caused by birds or codling moth caterpillars are often a point of entry for spores of the brown rot fungus. The infected fruits turn brown, the flesh becomes soft and quickly decays. They become covered with concentric rings of buff or whitish grey cushions of fungus spores. The disease spreads by contact. Affected fruits may fall, but eventually they shrivel and dry up.

The fungus can also enter shoots causing small cankers and dieback, and these should be cut off and burnt.

Treatment. Affected fruit, whether on the tree, on the ground, or in store, should be removed and burned as soon as it is noticed. The disease cannot be controlled effectively by spraying with a fungicide, although there is some evidence that spraying with thiophanate-methyl in mid-August and

Fig. 10.
Apples infected by
brown rot fungus.

early September, before picking the fruit, will reduce the rotting of apples in store.

Bruised or damaged fruits should not be stored and this applies even to those from which the stalk may have been torn off during picking. Store only dry fruit and examine it at regular intervals to remove rotting apples.

To keep the store clean, wash shelves or trays with soda and hot water or soak them in a 2% solution of formaldehyde (1 pint in 6 gal. water) for ½ hour at the end of the season.

Honey fungus. This root parasite frequently brings about the sudden death of fruit trees. White fan-shaped growths of fungus develop beneath the bark of the roots and the trunk of the tree at and just above ground level. Brownish black root-like structures known as rhizomorphs, may be found growing on diseased roots; these grow out through the soil and spread the disease. In autumn honey coloured toadstools sometimes appear at the base of the dying tree.

Treatment. Dead and dying trees should be dug out together with as many roots as possible and burnt. The soil should be sterilised with a 2% solution of formaldehyde (1 pint in 6 gal. water) applied at the rate of 5 gal. to the square yard, or the soil should be changed before replanting. A proprietary product containing a phenolic emulsion could also be used.

Papery bark. Bark becomes papery thin and peels off in a pale brown sheet. Many shoots may be affected and frequent dieback occurs following the girdling of shoots. The cause is unsuitable cultural conditions, especially waterlogged soil.

Treatment. Cut out all dead shoots. On the main trunk, remove dead and rotting tissues beneath peeling bark. Cover all wounds with a fungicidal tree paint. Improve the cultural conditions to prevent the trouble from persisting.

Fig. 11. An apple branch showing signs of papery bark.

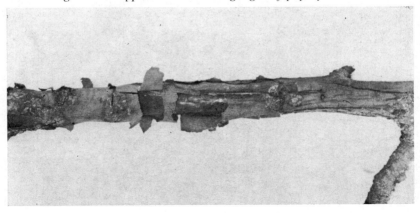

Fig. 12. The brown sunken spots associated with bitter pit.

Bitter pit. This physiological disorder affects only the fruit, producing slightly sunken pits on the surface of the skin and small brown areas of tissue immediately beneath the pits and scattered throughout the flesh. The symptoms may appear while the fruit is still on the tree, but most pitting develops during storage. Where the brown patches appear near the skin they can be removed by peeling the fruit, but in severe cases, the apples become almost inedible as the affected flesh has a bitter taste.

Bitter pit is due to a deficiency of calcium and/or a too high concentration of potassium or magnesium within the fruits, but its incidence is influenced by various factors. Thus it appears to be connected with a shortage of water at critical times and is generally worse in seasons when there are wide fluctuations in rainfall and temperature. It is more common on young, vigorously growing trees, especially those fed heavily with nitrogenous fertilizers, but it can also develop on fairly old trees, the trouble being most prevalent in large fruits and those which are immature when picked.

Treatment. Watering during dry periods will help to prevent bitter pit, and also mulching, but straw should not be used as this could aggravate the trouble. Incidence can be greatly reduced by spraying with hydrated calcium nitrate (at the rate of $\frac{1}{2}$ lb in 5 gal. water) starting in mid-June and repeating at least three times at 3-week intervals. The cultivars 'Cox's Orange Pippin', 'Crispin', 'Discovery' and 'Bramley's Seedling' are sensitive to calcium nitrate and on these, (and also on other cultivars if desired) flaked calcium chloride should be used in mid-June at $3\frac{1}{4}$ oz in 5 gal. water, repeating 10 days later at the same concentration, then three weeks later at $6\frac{1}{2}$ oz in 5 gal., repeating after another three weeks.

Apples may also be affected by Silver leaf (see page 24), and Fireblight (see page 20).

17

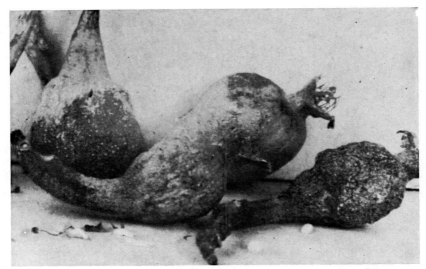

Fig. 13. Damaged pear fruitlets and the grubs of pear midge.

!Pears

Aphids. Several types occur on pear but the most serious is the pear-bedstraw aphid. It is covered with a mealy grey powder and in severe infestations the whole tree may be smothered, including the trunk. They cause severe leaf curl and fouling with honeydew and sooty mould before migrating to bedstraw for the summer.

Another species, the pear-coltsfoot aphid, causes the leaves to become red and folded along the mid-rib.

Treatment. The overwintering eggs can be destroyed by thoroughly spraying the tree during the dormant period with a tar-oil winter wash. A systemic insecticide applied shortly after petal fall will deal with any aphids that survived the winter treatment.

Pear midge. This is a localised pest which is inclined to affect the same tree year after year. The young fruitlets fail to develop and soon blacken and fall from the tree. Inside each affected fruitlet are numerous small maggots, up to 3–4 mm long, which later enter the soil, where they pupate. In spring the adult midges emerge and lay eggs on the pear blossom.

Treatment. This pest can be reduced by cultivating under the trees. The maggots spend about ten months of the year in the soil and cultivation will expose the maggots to predators and to the weather. Collecting and burning affected fruitlets will also destroy some of the maggots before they can reach the soil. If pear midge is a regular pest, the tree should be sprayed with BHC at the white bud stage in order to kill the adult midges.

Fig. 14. Damage to pear leaves by blister mites.

Pear leaf blister mite. This is a microscopic pest which overwinters beneath the bud scales and spends the summer months inside the leaves. Leaves which have been attacked by blister mites develop numerous pustules from April onwards. At first they are pale green or pink but they turn brown later and are most obvious by midsummer.

Treatment. If only a few leaves are affected, they are best picked off and burned. When an infestation is likely to be severe, spray thoroughly with 5% lime-sulphur at the end of March as the buds start to open. This will kill the mites as they leave their overwintering sites.

Pear sucker is related to apple sucker and it infests blossom buds and causes extensive fouling of the foliage with honeydew and sooty moulds. Infestations start by the end of March and may persist into late summer, so affecting fruit development and sometimes causing premature leaf-fall. Many adult pear suckers overwinter in hedgerows and similar protected situations and move on to pear trees from about mid-March to lay eggs. These soon hatch and the young suckers start feeding on the buds. Further generations of adults develop during the summer and the last adults mature in October.

Treatment. Winter sprays are not very effective against this pest as the the overwintering adults are not on the trees when such sprays are applied. One of the insecticides used against aphids will control this pest and it should be applied three weeks after petal fall.

Caterpillars. Looper and other types of caterpillar (see p. 10) can be controlled by spraying with trichlorphon whenever young ones are seen.

19

Fig. 15. A small canker of a branch (left) originating from an infection by fireblight of summer blossom at the end of a shoot.

Fireblight is caused by bacteria entering through the flowers. Normally infection in Britain occurs only through late spring or summer blossoms. The bacteria then spread down from the spurs and lateral twigs into the main branches. The shoots die back and the leaves become brown and withered but do not fall.

Cankers develop at the base of diseased tissues and they remain dormant during the winter. In spring the cankers become active again and droplets containing bacteria ooze out of them as the first flowers open. The bacteria are carried to the flowers by rain and insects.

Treatment. Fireblight is a notifiable disease and when a tree is affected by it the owner is obliged by law to notify the local representative of the Ministry of Agriculture. Full instructions will then be given as to how the diseased tree should be treated. This usually involves cutting out diseased wood to a point about two feet below the apparently affected tissues. All pruning tools must be disinfected after use.

Pear scab. Although similar to apple scab, pear scab is caused by a different fungus. The fruits develop blackish scabs, and in severe cases may crack. On the leaves the fungus produces olive green blotches and affected shoots become blistered and scabby.

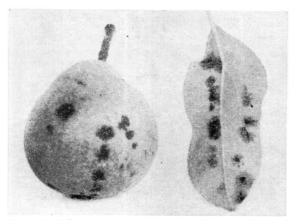

Fig. 16. left, scab damage on pear fruit and leaf; right, twig lesions caused by pear scab.

Treatment. Rake up and burn diseased leaves and cut out scabby shoots in the winter. Spray with captan or thiram (but not on fruit for preserving) at green cluster, white bud, petal fall and fruitlet stages and repeat at 14-day intervals; *or* with benomyl at bud burst, repeating at 21-day intervals; *or* use thiophanate-methyl at bud burst and repeat at 14-day intervals, but too frequent use of the latter two fungicides may cause the development of tolerant strains of the fungus.

In seasons when the weather conditions favour the development of scab it may be necessary to continue spraying until late July, whichever fungicide is used.

Lime sulphur will also control pear scab but it causes some damage and should only be applied before flowering (i.e. at green cluster and white bud). 'Doyenné du Comice' should not be sprayed at all with lime sulphur.

Pear stony pit virus. In Britain, this virus produces symptoms only on fruits, which are pitted and deformed at maturity and have patches of dead stony cells within the tissues. Such pears are inedible and in severe cases are so woody that they are difficult to cut. The disease usually first appears on the fruit of one branch of an affected tree, but gradually over the years all the fruit becomes pitted and distorted.

Stony pit virus can be transmitted by grafting and budding but other methods of transmission are not yet known, although it can spread slowly to other pear trees. This disease is most likely to be found on very old (and probably un-named) trees in gardens.

Treatment. There is no treatment for an affected tree and it should be destroyed.

21

Fig. 19.
A bad infestation by
mealy plum aphid.

Plum sawfly is one of the most important pests attacking the fruit. The caterpillars make conspicuous black, messy holes in the fruits which mostly drop from the trees. The adults emerge from cocoons in the soil in April. They fly onto the flowers and lay eggs which hatch two weeks later. The caterpillars then tunnel into the fruitlets. 'Czar', 'Belle de Louvain' and 'Victoria' plums are the most susceptible.

Treatment. Spraying with dimethoate, formothion or BHC about a week after the petals have fallen will kill the young caterpillars before they can tunnel into the fruitlets. Cultivating the soil under infested trees may reduce the numbers of pupae surviving from one season to the next.

Fruit tree red spider mite is sometimes troublesome on plums. See p. 11.

Silver leaf. Any fruit tree may suffer from silver leaf. Plums and in particular 'Victoria' are very susceptible to this disease. It shows as silvering of leaves which may become brown in severe cases. There is progressive dieback of affected branches and fairly small fruiting bodies of the fungus develop on the dead wood. These are purplish when fresh but become white or brown with age and they are variable in shape, being either bracket-shaped or overlapping like tiles or even flat on the branches. When an

24

Fig. 20.
Young plum fruitlets
attacked by
plum sawfly.

affected branch (at least an inch in diameter) is cut off, a brown or purplish stain can be seen in the cross section. (Fig. 21)

Silver leaf is caused by a wound parasite that enters through wounds such as pruning snags. The fungus produces a toxic substance which passes upwards in the sap and causes the leaves to become silvered.

Treatment. Destroy the tree only if fruiting bodies of the fungus appear on the main trunk. All dead branches must be cut back to about 6 inches behind the point where the stain in the inner tissues ceases. Paint the wound with a good fungicidal paint. Tools must be sterilised before and after use.

False silver leaf is a physiological disorder which is probably more common than true silver leaf and is often confused with it. The foliage again shows a silvery discoloration but most leaves on the tree are affected at the same time and there is little or no dieback. No stain is produced in the inner tissues of branches bearing silvered leaves. The discolouration of the foliage is due to malnutrition and/or an irregular supply of moisture in the soil.

Treatment. If the foliage becomes discoloured fairly early in the season, it is worthwhile trying the effect of spraying with a foliar feed. No other

Fig. 21. left, a healthy (top) and diseased (below) plum twig; right the characteristic stain in the trunk of a silver-leaf-infected plum.

treatment can be carried out during the summer but the following season the tree should be fed, mulched and watered as necessary. If any branches do die back, they should be cut back to clean living tissues and the wounds painted with a fungicidal paint.

Plum rust. Small bright yellow spots appear on the upper surface of an affected leaf. On the underside of each spot is a pustule bearing pale brown or orange spores and later, dark brown or blackish spores. Diseased leaves may turn yellow before falling prematurely. It is only troublesome on weak trees.

Treatment. The fallen leaves should be raked up and burnt. The following season the tree should be fed, mulched and watered as necessary and this treatment should prevent a recurrence of the disease. Should rust appear again, however, thiram can be used at the first signs of the trouble but may not check the disease completely.

Bacterial canker is now considered to be the most serious disease of plums. Brown rounded spots develop on the leaves. The tissues then fall away leaving holes so that, in severe cases, they look as though they had been eaten by caterpillars. Elongated flattened cankers exuding copious gum then develop on the branches. The following spring the buds of a severely infected branch either do not open or else produce small yellow leaves which become narrow and curled before withering and dying. The whole

26

branch then dies back. In other cases, the bacteria die out in the cankers, and no further symptoms are seen.

Treatment. Remove badly cankered branches and dead wood and paint all wounds with a good protective paint. As the bacteria live on the leaves during the summer, some control can be achieved by spraying three times with bordeaux mixture, starting in mid-August, with the second spray in mid-September and the final one in mid-October.

Shothole. Numerous brown rounded spots develop on the leaves. The tissues fall away leaving holes in the leaves so that in severe cases they look as if they have been eaten by caterpillars. No symptoms develop on the shoots. The fungus only attacks weak trees.

Treatment. The best method of preventing this disease is to feed, mulch

Fig. 22. Peach leaf showing typical symptoms of shothole which also affects plums.

and water trees so that they keep growing well. Once the disease has appeared, applications of a foliar feed during the summer may be beneficial, and the following spring the tree should be fed by soil applications. If the trouble occurs again, applications of a copper fungicide can be given, using it at half-strength during the summer and at full strength just before the leaves fall in the autumn.

Plums are also affected by Brown rot (see page 15) and Honey fungus (see page 16).

Cherries

Cherry blackfly is the only aphid that regularly attacks cherries. Eggs are laid near the buds on the twigs in autumn and hatch by the end of March. Dense colonies of black aphids develop at the tips of shoots and the leaves become distorted. Growth is checked and infested shoots may die back later.

Treatment. Spraying thoroughly with 5% tar-oil in December or early January kills overwintering eggs and malathion, or one of the systemics, such as dimethoate or formothion, sprayed just before flowering will kill off the young aphids before they become established.

27

Tan bark. The symptoms of this physiological disorder appear on one-year shoots or sometimes the main trunk of cherry trees. The breathing pores burst open and the bark splits, causing the outer layers to peel off exposing a brown rust-like tan-coloured powder which is made up of masses of dead plant cells. This trouble can occur where the root action is very vigorous, but it is also often due to wet soil conditions.

Treatment. Apart from draining the soil, if required, no treatment is necessary unless much of the bark has peeled. The loose bark should then be removed and also the dead tissues beneath to leave a clean wound which must be painted with a protective paint.

Cherries are also affected by: Winter moth (p. 10), Birds (p. 7), Silver leaf (see page 24), Shothole (see page 27), Bacterial canker (see page 26), and False silver leaf (see page 25), Brown rot (which is particularly trouble-some on acid cherries although it is not worthwhile spraying against it, p. 15) and Honey fungus (p. 16).

Peaches and Nectarines

Glasshouse red spider mite is troublesome under glass and on warm walls out of doors. Colonies of mites build up from about the end of April and feed on the undersides of the leaves. The first symptoms show as a fine light speckling which gradually extends over the leaf surface. The leaves become discoloured and may eventually bronze, wither and die.

Treatment. This pest is difficult to control. Few chemicals have any marked effect on it and the use of ineffective chemicals may result in an increase in mite numbers. Thorough spraying with malathion, dimethoate, formothion or derris early in the season may control low populations but repeat applications may be necessary later in the season. If chemical control fails biological control, based on a predacious mite, *Phytoseiulus persimilis*, should be considered.*

Aphids

As on other fruits aphids, particularly peach aphid, attack young shoots in the spring. They can be controlled by a spray of tar-oil in December, or malathion, dimethoate or formothion applied just before flowering.

Peach leaf curl is the commonest and most troublesome disease of peaches and nectarines (and also almonds). Reddish blisters develop on the leaves and later they swell up. The disease is caused by a fungus which eventually produces masses of spores on the surface of each swollen leaf giving it a whitish appearance. Many leaves may be affected and they fall very early

* Further information on this technique can be obtained from the R.H.S. Garden, Wisley, and supplies of the predatory mite are usually available from Wisley during the summer months. The predator will not control fruit tree red spider mite.

Fig. 23.
Two of the lower
leaves on this
shoot are badly
affected by
peach leaf curl.

in the season thus weakening the tree. The fungus dies out in the fallen leaves so does not overwinter on these on the ground.

Treatment. It is essential to spray early so as to prevent the entrance of the germinating spores into the buds. Lime sulphur (3%), bordeaux mixture and liquid copper can all be used, and the trees should be sprayed in January or February, with a second spray 10 to 14 days later. Another spray in autumn just before the leaves fall will complete the control. Spraying diseased trees during the spring and summer is not worthwhile.

Peach mildew. In some gardens, powdery mildew can be troublesome on both peaches and nectarines. It is due to a fungus which overwinters both within the buds and as small blackish fruiting bodies on the shoots. In spring the infected buds burst to produce stunted and diseased leaves which are covered with a mealy coating of spores. During the summer the spores spread the disease to later developing leaves and shoots which also become coated with white powdery spores. Diseased leaves fall prematurely. The fruit can also be attacked, but the disease is not very obvious at first on peaches because of the fruit hairs, but on the maturing fruit it shows as brownish patches.

Mildew is likely to be most troublesome on a tree growing close to a wall where there is very little air movement and also on trees affected by too dry soil conditions.

Treatment. Mulching and watering before the soil dries out completely will help to prevent mildew. This is particularly necessary for wall trees as the soil at the base of a wall may remain dry during heavy rain if the prevailing wind is not in the right direction. It is also advisable to keep the training wires six inches, or more, away from the wall to ensure adequate air circulation through the branches.

Cut out all infected shoots to a point several inches below the apparently diseased tissues. Apply a sulphur fungicide at the first signs of the disease, repeating as necessary at fortnightly intervals.

Split stone

Peaches affected by split stone have a deeper suture than normal and are cracked at the stalk end, the hole sometimes being big enough for earwigs to enter. The stone of such a fruit is split in two and the kernel (if formed) rots.

One or more adverse factors can cause this physiological disorder. It sometimes occurs as a result of poor pollination and it can be due to lack of lime, general malnutrition or feeding after stoning has started. But the commonest cause is an irregular supply of moisture in the soil.

Fig. 24. Mildew on shoots and fruits of peaches.

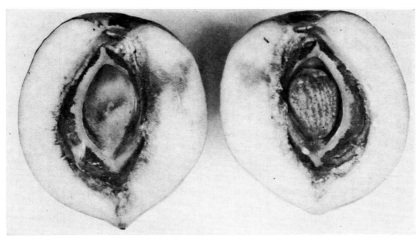

Fig. 25. Split stone of peaches.

Treatment. Split stone can be prevented in several ways. Hand pollination of the flowers, particularly if there are no bees about, can be done by passing from flower to flower a soft camel-hair brush tied on the end of a short bamboo cane. In dry weather gently syringe the flowers of trees on walls (and under glass) with tepid water about noon each day to help fruit setting. If the soil is acid it should be limed in the autumn. It is essential to mulch and water in dry periods so that the soil remains damp.

Silver leaf, (p. 24), Shothole (p. 27), Bacterial canker (p. 26), False silver leaf (p. 25), Brown rot (p. 15), and Honey fungus (p. 16) can also affect peaches and nectarines.

Apricots

Apricot dieback. Dieback of shoots and large branches is a common disorder of apricots. It is sometimes due to adverse cultural conditions such as drought, waterlogging or malnutrition, and occasionally it is caused by late spring frosts. In most cases, however, it is due to one or more fungi which enter the tissues through a small wound and then kill the branch.

Treatment. It is essential to ensure that apricots are well fed, mulched and watered. All dead wood should be cut back to healthy living tissues and the wounds covered with a protective paint.

Silver leaf and Bacterial canker can also affect apricots. The symptoms and treatment are as described for plums (see pages 24 and 26).

31

Currants

Aphids. At least six different aphids attack currants. Some cause conspicuous red or yellow blistering of the leaves, others stunt and distort leaves and shoots. Persistent aphid attacks may affect growth.

Treatment. Currant bushes are easily sprayed and it is well worth applying a winter-wash of either tar-oil or DNOC/petroleum in January to kill off overwintering eggs. This should be done before the end of January, before the young buds start to grow.

If a winter-wash has not been used, the young aphids hatching in late March and early April can be controlled by spraying with malathion or with one of the systemic insecticides, such as dimethoate or formothion. These insecticides should not be used during flowering but further applications may be made after flowering has finished, providing the safe minimum periods between application and harvest are observed. For malathion this is at least four days and for dimethoate and formothion it is one week.

Capsid bugs. The leaves at the tips of the shoots often become tattered with many small holes as a result of feeding by this pest. They can be controlled as described under apples (p. 10).

Big bud gall mite. This is the most important pest of black currants as the mites cause the condition known as "big bud" and also transmit reversion which can lead to a reduction in cropping and general deterioration of infested plants. Red currants and gooseberries may also be attacked by this mite but the effects are never as serious as on black currants.

The extremely small mites invade the new buds between April and July. Many thousands live within each bud and attacked buds swell but fail to develop. These enlarged and rounded buds are particularly conspicuous during the winter and early spring and are easily distinguished from the narrow, pointed healthy buds. In the spring the mites leave the old "big buds" and migrate over the bush and this is the only time that they leave the protection of the buds. The mites are dispersed by means of wind, rain and insects, to which they may cling, and during this dispersal they may carry reversion from infected plants to healthy ones.

Treatment. Black currant bushes should be examined carefully in January, February and March and any enlarged buds picked off and burned, to kill the mites inside them before they disperse. In spring the bushes should be sprayed with $\frac{1}{2}$ to 1% lime-sulphur when the first flowers open, which is usually in early April, repeating this treatment after about three weeks. Some black currants such as 'Davison's Eight', 'Goliath' and 'Wellington XXX' are sensitive to lime-sulphur and should only be treated with $\frac{1}{2}\%$ lime-sulphur.

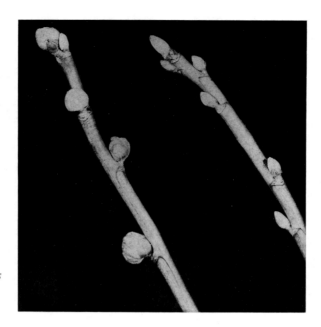

*Fig. 26. Currant
shoot on the left has
big buds; that on
the right is healthy.*

If bushes are affected by reversion (see below) to the point where yields are inadequate, they should be removed and replaced. Do not propagate from infected bushes but buy in healthy new stock and, if possible, plant on a new site.

Reversion. On affected plants mature leaves on the leafy basal shoots are narrow and have less than five pairs of veins on the main lobe. These symptoms are best seen in June or July. The flower buds on "reverted" plants are of a bright magenta colour and not dull and grey as on healthy bushes. Unfortunately these symptoms are difficult for an amateur to identify, particularly as in spring and late summer many of the young leaves on healthy bushes may look abnormal. Diseased bushes will, over the years cease to produce a good crop, but poor yields can also be due to frost damage, poor pollination or a deficiency of potassium. When bushes fail to give a satisfactory crop, therefore, specialist advice should be sought in June or July for an opinion on the reduction in yield.
Treatment. The first control is to buy new plants that are certified as being healthy. The second is to control big bud mite (see p. 32).

Once confirmed, badly diseased bushes should be destroyed, particularly if only individual plants are affected among otherwise healthy ones. If

Fig. 26. The shoot and leaf on the left are reverted; those on the right are healthy.

many bushes are infected, but are still producing a worthwhile crop, the bushes can be retained until they stop fruiting, when they should all be destroyed at the same time. New currant bushes of certified stock should, if possible, be planted on a fresh site.

Leaf spot. This disease can affect black, red and white currants. From May onwards, it shows on the leaves as small dark brown spots which coalesce until the leaf surfaces become completely brown. Affected leaves fall prematurely and so the crop in the following season is smaller. The disease is spread during the summer by means of air-borne spores, and during the winter the fungus remains dormant on the fallen leaves.
Treatment. Diseased leaves should be collected and burnt to destroy the overwintering stage of the fungus. If the disease becomes troublesome, spray the bushes after flowering with zineb or thiram, repeating at 10- to 14-day intervals to within a month of fruit picking. Thiophanate-methyl will also control this disease if applied as the first flowers open, repeating two or three times at fortnightly intervals. Benomyl can also be used, first at the grape stage (before the flowers open) and repeating three times at 14-day intervals. These various fungicides can also be used after harvest if necessary and copper fungicides can also be applied then. Diseased bushes should be well fed to help them overcome the loss of vigour due to the early leaf fall.

Coral spot. Red currants are very susceptible to infection by the coral spot fungus which enters through small wounds such as pruning snags and

34

Fig. 28.
Gooseberry
shoots showing
typical bird
damage.

causes dieback of branches or occasionally the complete death of a plant. The dead shoots bear numerous coral-red pustules or spores.

Treatment. Affected braches should be cut out to a point several inches below the apparently diseased tissues and all wounds should be painted with a protective paint. All infected shoots and woody debris in the garden, such as old pea sticks on which the fungus frequently lives as a saprophyte, should be burnt.

American gooseberry mildew. Black currant shoots are occasionally affected by this powdery mildew late in the season. Symptoms and treatment are essentially the same as for gooseberries (see page 36) but spraying need not start until the disease appears on the bushes.

Grey mould can cause currants to rot. See page 41.
Silver leaf. See page 24.
Honey fungus. See page 16. Can also affect currants.

Gooseberries

Aphids. Some species attack gooseberries. Symptoms and treatments are essentially the same as those described for currants (p 32).

Birds. Bullfinches can cause severe damage by eating buds during the winter and 'Leveller' seems to be especially susceptible. The only sure way of preventing such damage is to put netting over the bushes from about mid-November to mid-March. The ripening fruit will also require protection from birds in some areas.

Gooseberry sawfly. In April and May, gooseberry sawfly caterpillars start to feed on the foliage. They are green with black spots and they mostly attack two- and three-year-old bushes. They start feeding in the middle of the bush and work outwards, so that, unless a careful watch is maintained, the bush may be defoliated before the pest is noticed. Attacks may continue until the autumn and the caterpillars overwinter in cocoons in the soil.
Treatment. The caterpillars are conspicuous, and if noticed before they have done much damage, they can be removed by hand picking. If they have become established spray thoroughly with derris, malathion or nicotine, but do not spray during flowering. This treatment may be repeated whenever necessary but there is no point in applying insecticides if the caterpillars are not present.

American gooseberry mildew. This disease shows as a white powdery coating on the young leaves, shoots and especially the fruits. The white patches later turn brown and become felted. Badly affected shoots become distorted at the tips. It is most likely to be troublesome on overcrowded bushes where there is little circulation of air, and also on plants which have been grown too soft (i.e. have been given too much nitrogenous fertilizer).
Treatment. Prune the bushes regularly to keep them open, allowing good circulation of air through them as this will help to prevent the disease. Cut out and burn diseased shoots, preferably in late August or September. Do not apply too much nitrogen.

Fig. 29.
Gooseberry leaf and fruit infected by American mildew.

36

Treatment. Lime sulphur at $1\frac{1}{2}\%$ can be used (except on fruit to be pre-served to avoid possible taint, or on the sulphur-shy cultivars 'Careless', 'Early Sulphur', 'Freedom', 'Leveller', 'Lord Derby', 'Roaring Lion', 'Yellow Rough' and 'Golden Drop'). The first application should be given just before the flowers open, repeating at fruit set and 3 weeks later. On sulphur-shy gooseberries use dinocap starting before flowering and repeat as necessary until 1 to 2 weeks before harvesting. Benomyl and thiophanate-methyl may also control the disease if applied as the first flowers open, repeating two or three times at 14-day intervals. Any of these fungicides can be used after harvesting if necessary.

Gooseberries are also affected by Leaf spot (p.34), Honey fungus (p. 16) and Grey mould (p. 41). The latter disease, however, is more troublesome on the shoots causing them to die back. Such dead shoots should be cut out and all wounds covered with a protective paint.

Raspberries, Loganberries and Blackberries

Aphids. At least three different species of aphid attack these fruits. The direct damage caused by their feeding is not particularly harmful but they can cause considerable harm indirectly by transmitting some virus diseases. In the garden the high level of aphid control that would eliminate any risk of viruses spreading is unlikely to be achieved. Try to keep aphid infestations to a minimum by applying the control measures recommended for use against currant aphids (see p. 32). This will also control leafhoppers, which also transmit raspberry viruses.

Raspberry beetle. The grubs of this small beetle feed on the ripening fruit and are often noticed when the fruit is picked and prepared for eating. Eggs are laid in the blossoms by the beetles, and grubs may be seen in raspberry, loganberry or blackberry fruits at any time between early June and early September. When they have finished feeding the grubs pupate in cocoons in the soil.

Treatment. The raspberry beetle is a common pest throughout the country and in every hedgerow that contains blackberries there is probably a reserve from which cultivated fruits are infested. Chemical protection is therefore essential if the fruit is to be kept clean. Malathion or derris sprays will generally give good control if they are applied on the following occasions:

—On raspberries when the first pink fruit is seen.

—On loganberries, when flowering is almost over and repeated when the first fruit is starting to colour. (The same treatment may be given to raspberries if top quality dessert fruit is required).

—On blackberries, as the first flowers open.

Spray in the late evening to avoid killing bees on the flowers.

Spur blight is one of the most troublesome diseases of raspberries in gardens and it can also affect loganberries. It is always worse where canes are overcrowded. It is caused by a fungus which infects the new canes in June.

Symptoms do not usually appear until August when dark purplish blotches develop around the nodes on the canes. The blotches enlarge, turn silver and then become studded with minute black fruiting bodies of the fungus. The buds at the affected nodes wither and die or else produce shoots in the spring which soon die back.

Treatment. Superfluous young canes should be removed early and all diseased canes should be cut out and burnt as soon as the first symptoms are seen. The disease can be controlled by spraying with benomyl, dichofluanid, thiram or captan starting when the new canes are only a few inches high and repeating three or four times at fortnightly intervals. Thiophanate-methyl can also be used, giving the first application when the buds are $\frac{1}{2}$ inch long and repeating at 14-day intervals until the end of flowering. Bordeaux mixture or liquid copper can also be applied to canes when the buds are $\frac{1}{2}$in. long (at bud burst) and again when the tips of the flowers are just showing white.

Cane spot. Raspberries, loganberries and hybrid berries can all be attacked by cane spot but it is rarely troublesome on blackberries. The first signs

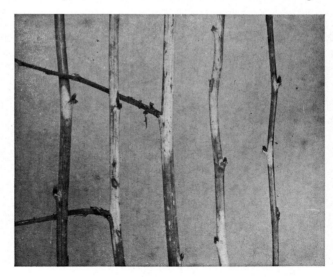

*Fig. 30.
Raspberry
spur blight.*

38

of the disease appear in May or June when small purple spots develop on the canes. As the spots enlarge they become elliptical, up to $\frac{1}{4}$ inch long, whitish with a purple border, and eventually split to form shallow pits or small cankers which give the canes a rough and cracked appearance. In severe cases the tips of the shoots die back. Small spots may also appear on the leaves, and infected fruits become distorted. The fungus which causes the disease overwinters on the canes and re-infection takes place from May until October.

Treatment. Badly spotted canes should be cut out and burnt in the autumn to destroy the overwintering stage of the fungus.

Control the disease on raspberries by applying 5% lime sulphur at bud burst (i.e. when the buds are about $\frac{1}{2}$ inch long) and $2\frac{1}{2}\%$ lime sulphur immediately pre-blossom: *or* liquid copper or thiram (except on fruit to be preserved) at bud burst and pre-blossom: *or* benomyl or thiophanate-methyl at bud burst, repeating at 14-day intervals to the end of blossoming.

On loganberries use bordeaux mixture or liquid copper or thiram (except on fruit to be preserved) at pre-blossom and as soon as the fruit has set.

Cane blight. The raspberries 'Norfolk Giant' and 'Lloyd George' are very susceptible to this disease and it can also affect other types of raspberry but not loganberries, blackberries or hybrid berries. The disease is caused by a fungus which enters the bases of the canes, usually through frost cracks.

A dark area develops on the canes just above ground level and they become so brittle that they are easily snapped off. During the summer, the leaves wither on infected fruiting canes. The soil can become contaminated with the fungus once it has entered canes at soil level.

Treatment. All diseased canes should be cut hard back to below soil level and burnt. The pruning knife or secateurs should be disinfected immediately afterwards. Apply bordeaux mixture or another copper fungicide as the new canes grow.

Crown gall. Raspberries and blackberries can be affected by crown gall which shows as a walnut-sized gall at ground level or as a chain of small galls higher up the canes. It is caused by a bacterium which enters through wounds and which persists and spreads in wet soils.

Treatment. Destroy severely diseased canes and replant on a fresh site, where drainage is better.

Virus diseases. Raspberries are particularly susceptible to virus infection as they can be affected by several different viruses, but loganberries may also be affected by certain virus diseases. The viruses may be transmitted by aphids or eelworms or may be seed- or pollen-borne. In general the

Fig. 31.
Leaf mottling on
raspberry leaves
caused by virus.

symptoms of virus infection are yellow mottling or blotching and distortion of the leaves. Growth is poor and the yield is reduced in severely affected canes.

Raspberries, hybrid berries and in particular, blackberries, can also be affected by a mycoplasma which is an organism having certain characteristics in common with both viruses and bacteria, but the disease it causes is very similar to a virus disease because affected plants are extremely stunted and the crop is reduced as few flowers are produced. The disease is transmitted by leafhoppers.

Treatment. There is no cure for virus or mycoplasma infection and affected stools should be removed and burnt. Only canes certified to be healthy should be bought for replanting and they should be planted on a fresh site at least 50 feet away from any old canes which are to be retained. If a fresh site is not available, all the old canes should be destroyed at the same time and the soil changed completely to a depth of $1\frac{1}{2}$ feet before

40

replanting. If possible, cane fruits should be planted well away from hedges which may harbour the eelworm vectors of some of the viruses.

Grey mould (see below) and Honey fungus (see page 16) frequently affect raspberries.

Strawberries

Slugs. Various species of slug eat the ripening fruits. Slugs are encouraged by a high organic content in the soil and the use of straw or other mulches probably attracts them.
Treatment. It is best to reduce slug populations before fruiting and good hygiene during autumn and winter will help. Slug baits, especially those based on methiocarb, may be used to reduce the local population. Chemicals should not be used after the plants have started to fruit.

Strawberry beetles. These black, shiny beetles also eat the fruit and particularly the seeds. Like slugs, they are favoured by accumulations of plant debris which provide them with cover. The numbers can be reduced by good hygiene by keeping down weeds and removing dead leaves and by trapping beetles in pit-fall traps made by sinking jam-jars level with the soil surface. Methiocarb slug pellets are also effective against them.

Birds and mammals. Birds, such as blackbirds, and mammals, such as squirrels, will take ripening fruits and the only way to prevent this is to cover the plants with netting.

Aphids, Eelworms, Mites. Various species of aphid, eelworm and mite attack strawberries and can produce similar symptoms, usually in the form of stunted growth and distorted leaves. As correct treatment depends on correct diagnosis it is best to seek expert advice in doubtful cases. Aphids infesting young growth are controlled by spraying with malathion or nicotine. Control of mites and eelworms may prove difficult.

Grey mould can result in serious losses of fruit in a wet summer when affected berries rot and become covered with a brownish grey fluffy growth. It is due to a fungus which enters the flowers but the disease is not usually seen on the fruit until it is beginning to ripen. Once grey mould has appeared, however, it can spread very rapidly by contact between diseased and healthy fruits. There are always spores of the fungus in the air as it can attack every type of plant, and it is frequently found on dying weeds and also on thick straw mulches.
Treatment. When grey mould first appears on strawberry fruits it is too late to apply any chemical. But infection by contact can be reduced by removing and destroying all infected fruits when harvesting. To prevent infection, a fungicide must be applied as soon as the first flowers open.

Benomyl and thiophanate-methyl give good control if applied three times at 10- to 14-day intervals. Another effective fungicide is dichlofluanid applied four times at 10-day intervals, but do not use this on strawberries under glass or polythene: it does sometimes scorch the foliage of less well-known strawberry cultivars. Thiram and captan can also be used except on fruit for preserving.

Strawberry mildew. The disease shows as purplish patches on the leaves which curl upwards exposing the lower surfaces. These have a greyish appearance due to the development of white fungal spores which spread the disease to other leaves, flowers and berries, which lose their shine or may even become shrivelled. The fungus overwinters on old green leaves.

'Royal Sovereign' and 'Cambridge Vigour' are very susceptible to powdery mildew but 'Cambridge Favourite' is resistant.

Treatment. Dust with sulphur or spray just before flowering with $1\frac{1}{2}\%$ lime sulphur, or with dinocap, repeating at 10- to 14-day intervals until 1 or 2 weeks before harvest, except on fruit to be preserved. Benomyl and thiophanate-methyl will also control mildew if applied three times at 14-day intervals from early flower stage through to white fruit. After harvesting, cut off the old leaves or spray again with one of these fungicides.

Leaf spots and leaf blotch. Several different fungi can produce small circular purple, red or white spots on strawberry leaves, but in general, the symptoms are only slight, occurring most commonly on old leaves at the end of the season. Occasionally, however, severe infections occur when the spots coalesce and the leaves wither and disintegrate.

A more serious disease, which is known as strawberry leaf blotch, appears to be on the increase in gardens. Large brown blotches with a purplish border and surrounded by a yellowish zone may develop on the leaves but blackening and rotting of the leaf- and flower-stalks can occur with subsequent withering of the fruits and death of the leaves without any blotches first appearing.

Treatment. Any leaves showing severe spotting should be removed and burnt. If the trouble persists and there is much withering of leaves and fruit, the plants should be sprayed with a fungicide the following season. Zineb at a rate of $\frac{3}{4}$ oz in $2\frac{1}{2}$ gal. of water, can be applied in spring just after growth starts and again a fortnight later; or dichlofluanid can be used at 2 to 3 weeks before flowering and again 10 to 14 days later.

Virus diseases. Strawberries can be affected by a number of different viruses which cause stunting or the complete collapse of a plant. Diseased plants give little if any fruit.

Fig. 32. Strawberry virus; an infected plant on the left and a healthy one on the right.

The symptoms are most obvious in April and September when diseased plants show leaf symptoms such as yellow edges, yellow or purplish mottling or blotching, and dwarfing and puckering. Most of the viruses which are troublesome on strawberries are spread by aphids.

A somewhat similar disease is *green petal* which causes the flowers to produce green petals, prevents fruit from ripening and results in the collapse of the plant. The symptoms are most obvious in midsummer when old leaves turn red and young leaves become yellow, before the plants die. This trouble is transmitted by leaf-hoppers, and is caused by a mycoplasma.

Treatment. All plants showing any of the above symptoms should be destroyed and no runners taken from them. If a stock is severely affected, all the plants should be burnt. Only plants certified to be healthy should be bought, and if possible, they should be planted on a fresh site well away from hedges. If the new strawberries have to be planted in the same position as those which have been removed, the soil should be changed completely (to a depth of 1 foot) before replanting.

Vines

Grape mildew is most troublesome on vines growing in cold greenhouses, but in some seasons it can cause severe damage to outdoor grapes, particularly on plants against walls. It shows on the leaves, shoots and fruits as a white powdery deposit. The skins of diseased fruits may become discoloured, but more frequently, the berries split and are then often

43

affected by secondary fungi such as grey mould, which cause extensive rotting.

Treatment. Spray with dinocap at the first signs of trouble and repeat as necessary. Under glass this fungicide can be used regularly as a smoke formulation. Like all powdery mildews this disease will flourish in stagnant air conditions, so overcrowding of the shoots and leaves must be avoided. An adequate supply of moisture to the plant must be maintained by mulching and watering. When the disease appears on vines under glass, ventilate the greenhouse well, and, if possible, provide some form of temporary heat in cold and dull weather.

Honey fungus frequently kills vines, including plants growing in green-houses.

Magnesium deficiency. Vines seem to be very susceptible to this trouble which shows as a yellowish orange discolouration between the veins. Later the affected areas turn brown.

Treatment. Spray with $\frac{1}{2}$ lb. magnesium sulphate in $2\frac{1}{2}$ gal. water, plus a spreader, two or three applications being given at fortnightly intervals.

Spray Calendar

The timing of spray application for control of pests and diseases is based on the stages of bud development. Dates for these stages can only be approximate because the rate of growth will vary according to locality and season.

In January

Spray dormant fruit trees, canes and bushes with tar oil, to control over-wintering aphid and sucker eggs—if this was not done in December.

Inspect apples and pears for canker and treat where necessary.

Spray peaches and nectarines against peach leaf curl. Repeat in 10 to 14 days.

In February

Complete tar oil spraying on apples and pears. As an alternative to tar oil use DNOC-petroleum wash on apples in late February if there is evidence of heavy infestation of red spider mite.

Spray peaches and nectarines against peach leaf curl. Repeat in 10 to 14 days. Inspect apple and pear trees for canker and treat if necessary. Spray severely infected trees. Protect all large pruning cuts.

In March

If a winter wash has been applied control of aphid, sucker and scale insects may not be necessary, but if aphids are present spray with a systemic insecticide. Spray apples against scab at bud burst and again at green cluster. Spray against aphids, apple sucker, capsid and caterpillars.

Spray pears against scab at bud burst, green bud and white bud (10 to 14 day intervals). Control aphids and cater-pillars, and if necessary pear midge and pear leaf blister mite.

Spray plums, cherries and damsons against aphids and caterpillars at bud burst to white bud.

Spray black currants at the grape stage to control leaf spot, mildew, grey mould, and against gall mite (big bud) at the late grape stage and repeat after 3 weeks. Spray against aphids if present.

Spray gooseberries against gooseberry mildew just before the flowers open; repeat at fruit set and 14 days later.

Spray cane fruits against spur blight and cane spot, at bud burst, and against aphids if necessary.

In April

Do not use insecticides during flowering of any crop as bees and other pollinating insects will be killed.

Continue scab control on apples and if not already done, control aphids, apple sucker, capsid, caterpillars and woolly aphids before open blossom. Start spraying against mildew at green cluster if using thiophanate-methyl or benomyl, or at pink bud if using dinocap. Continue with scab control on pears (white bud to full bloom); also control aphids and caterpillars, and, if necessary, pear midge and pear leaf blister mite before open blossom.

Spray peaches and nectarines at petal-fall against red spider mite.

Control sawfly on plums at cot split (about 8 days after flowering) if necessary.

Control aphids and caterpillars on cherries before open blossom.

Apply the second spray to black currants against gall mite. Spray for aphids if present. Start spraying against leaf spot. Continue control of gooseberry mildew on gooseberries. Control gooseberry sawfly. Start spraying against leaf spot, if necessary.

Inspect strawberries for aphids and spray if necessary. Spray at first open flower against grey mould (botrytis) on plants under glass or polythene (but do not use dichlofluanid).

Spray raspberries against spur blight if necessary.

In May

Never apply insecticides during flowering as they will kill bees and other pollinating insects.

Control apply sawfly and capsid at petal-fall, also red spider mite if present. Continue scab and mildew control.

Spray against pear scab at petal-fall and repeat as necessary.

Control sawfly on plums at cot split, if necessary.

Continue leaf spot control on currants and start control of gooseberry mildew if necessary.

Spray gooseberries against mildew and leaf spot. Control sawfly caterpillars if present.

Spray raspberries before flowering against cane spot and spur blight. Spray against grey mould.

Spray blackberries, etc. against cane spot in mid-May, and against spur blight if necessary.

Spray or dust strawberries immediately before flowering against mildew; repeat 10 to 14 days later. When the first flowers open spray against grey mould. Control slugs if present.

In June

Continue to spray apples regularly against scab and mildew. Inspect for red spider mite and spray if present. Spray against codling moth about mid-June and again 3 weeks later. Spray against bitter pit in mid-June if necessary.

Inspect stone fruits for red spider mites and aphids and spray if present.

Continue leafspot and gooseberry mildew control on currants.

Inspect gooseberries for caterpillars of gooseberry sawfly and magpie moth and control if present. Spray once or twice against leaf spot after harvest if necessary. Control grey mould and mildew on strawberries. Control slugs if present.

Spray cane fruits against spur blight and cane spot immediately before flowering and against grey mould as the first flowers open. Spray against raspberry beetle when most blossom is over and again after 14 days.

In July

Apply a second spray to apples against codling moth caterpillars (3 weeks after the first). Continue regular sprays against mildew until mid-July. Continue spraying against bitter pit if necessary.

Check whether scab is present and if necessary continue spraying until mid-July. Inspect for woolly aphid and treat if found.

Apply second spray to blackberries,

etc. against raspberry beetle 14 days after the first.

On currants and gooseberries continue to control leaf spot and gooseberry mildew after harvesting.

In August

On apple give last spray against bitter pit. Spray mid-August to reduce losses from brown rot in store.

Spray stone fruits after harvest but not before mid-August against bacterial canker if necessary.

Continue with leaf spot control on gooseberries and currants after harvesting, and with control of gooseberry mildew if necessary.

In September

Look for canker on apple and treat. Spray in early September to check brown rot in store.

Apply a second spray to stone fruits against bacterial canker if necessary.

In October

Check apple trees for canker and control if necessary; spray severely affected trees just before leaf-fall.

Greaseband apple and cherry trees on which spraying is not feasible. Apply final spray at leaf fall to stone fruits against bacterial canker. Spray peaches and nectarines against peach leaf curl; this will suffice as the mid-October spray against bacterial canker if a copper fungicide is used.

In November

Net or cotton all fruits (but especially gooseberries and plums) against birds where possible.

Inspect apples and pears for canker and treat: spray severely infected trees just before leaf-fall and at 50% leaf-fall.

Spray peaches and nectarines against peach leaf curl just before leaf-fall if not already done.

In December

Start spraying of dormant tree, bush and cane fruits with tar oil winter wash to control aphids, sucker and scale insects. (If red spider mite has been troublesome see February.)

Complete the spraying of stone fruits by the end of the month.

Inspect apples and pears for canker and treat where necessary.

46

Fig. 1

Fig. 2

Fig. 3

Fig. 4

Fig. 5

Stages of apple development
Fig. 1. *Bud burst stage*
Fig. 2. *Green cluster stage*
Fig. 3. *Pink bud stage*
Fig. 4. *Petal fall stage*
Fig. 5. *Fruitlet stage*

Index